true green

true green

100 everyday ways you can contribute to a healthier planet

Kim McKay and Jenny Bonnin

NATIONAL GEOGRAPHIC

WASHINGTON, D.C.

Published by the National Geographic Society
1145 17th Street, N.W., Washington, D.C. 20036-4688

First published in Australia by ABC Books for the
Australian Broadcasting Corporation
GPO Box 994 Sydney NSW 2001

Design, layout and select images by Marian Kyte
Edited by Tim Wallace

ISBN 10: 1-4262-0113-3
ISBN 13: 978-1-4262-0113-4

Library of Congress Cataloging-in-Publication Data available upon request.

A percentage of proceeds from the sale of *True Green*
benefits Clean Up the World.

True Green is a pending Trademark of True Green (Global) Pty Ltd.

One of the world's largest nonprofit scientific and educational organizations, the National Geographic Society was founded in 1888 "for the increase and diffusion of geographic knowledge." Fulfilling this mission, the Society educates and inspires millions every day through its magazines, books, television programs, videos, maps and atlases, research grants, the National Geographic Bee, teacher workshops, and innovative classroom materials. The Society is supported through membership dues, charitable gifts, and income from the sale of its educational products. This support is vital to National Geographic's mission to increase global understanding and promote conservation of our planet through exploration, research, and education.

For more information, please call 1-800-NGS LINE (647-5463) or write to the following address:

National Geographic Society
1145 17th Street N.W.
Washington, DC 20036-4688 U.S.A.

Visit the Society's Web site at www.nationalgeographic.com/books

Printed in U.S.A. on recycled paper.

contents

foreword

Ian Kiernan, AO
Chairman & Founder, Clean Up the World

During the past 17 years I have been lucky enough to see firsthand just how much people around the world care for their environment. Whether it's in the United States or Uganda, Australia or Angola, Poland or the Pacific Islands, or in China or Chile, ordinary people have made extraordinary efforts through the Clean Up the World program to improve their environment. Millions of people from school children to Nobel Peace Prize recipient Wangari Maathai have joined Clean Up the World, demonstrating that, the world over, we share the common goals of breathing clean air, drinking clean water, and living harmoniously with nature.

Sailing solo around the world in 1986-87 awakened in me a desire to do something positive to protect our environment. While sailing through the Sargasso Sea in the mid-Atlantic, I realized just how much impact we were having on our natural world. Where I should have seen only seabirds and the fabled golden seaweed, I saw all sorts of man-made durable plastic items, from buckets to combs littering the ocean.

Not long after, a group of friends and I started the Clean Up Australia and Clean Up the World campaigns. Becoming involved and taking positive action has also prompted this book, because we all face a major challenge over changes to our environment. The world's climate is changing more dramatically than we would have believed even a year ago. It's no longer a prediction. The evidence is real, and, unless we do something now, the effects will be significant and damaging.

The good news is that we can do something about our changing climate. This doesn't mean we have to sacrifice the lifestyles we enjoy now. I believe the changes necessary are easy to achieve and simple to do. I hope you will join with me in taking some of the steps outlined in *True Green*.

My tip: choose the top ten changes you'll make today, work your way up to more next week, and set new goals for the weeks after that. It's simple, it's fun, it's rewarding—because every action we take does make a difference.

introduction

Kim McKay and Jenny Bonnin

Every day we read in the newspapers or see on television reports about global warming and climate change. There seems to be a cascade of information and differing scientific opinions. What are we to believe and, more importantly, what can we do?

The vast majority of the world's leading scientists and experts agree that global warming is a reality. The question facing all of us is how to live with the effects of climate change and minimize its future impact.

We are already seeing the effects of global warming through extended droughts and intense hurricanes. We are witnessing, too, the disruption of wildlife habitats in places like the Arctic Circle.

Governments and corporations are investigating ways to decrease our dependency on fossil fuels, but there is no time for complacency. We need to look immediately at developing renewable energy sources like solar and wind power—of which the U.S. has an abundance—to feed our energy needs.

The solution lies with all of us.

When we helped establish Clean Up Australia and then Clean Up the World in the early 1990s, it was in response to a collective frustration evident in the local community. Why were our waterways choked with nondegradable consumer waste products? Why was raw sewage flowing freely onto our beaches?

We've seen firsthand how community action can change things—there are literally thousands of locations around the globe where communities have worked to clean up and fix up point sources of pollution. That same tide of public concern is evident again. We believe we're about to ride a new wave of community action in support of the environment.

True Green illustrates how everyone can make a positive difference to the environment by making simple lifestyle changes. Encourage your neighbors and friends to join you, and involve your family in adopting the practical and easy-to-manage suggestions included here—after all, it's their environment, their health, and their lifestyle that's at stake. You may not be able to do all 100 tips at first, but together you'll be making a real impact.

Have fun, let us know if you have other ideas, and make *True Green*'s and Clean Up's goals for a sustainable world part of your everyday life.

you *can* make a difference

The U.S. is the source of a quarter of the world's greenhouse gas emissions—hardly an auspicious position for a continent with so much to lose from global warming. The best we've done to date to rectify this is to limit the rate at which emissions are increasing. It's not just the fault of big business—after all, our affluent lifestyles are built on the economic benefits of cheap power from greenhouse-polluting coal and export dollars derived from energy-intensive primary industries. Households contribute almost one-fifth of America's greenhouse gases—more than 80 tons of toxic carbon dioxide per household per year—and much of that can be avoided through the easy steps outlined in this book. Get started today. Decide to follow these simple suggestions and immediately reduce your ecological footprint.

www.betruegreen.com
www.thegreenguide.com

Americans are often known as "the ultimate consumers," and for good reason. The United States is the source of a quarter of the world's greenhouse gas emissions, with the average U.S. household generating more than 80 tons of toxic carbon dioxide gases every year—that's 22.5 tons of greenhouse gases per person, per year. In addition, the average American family uses 102,000 gallons of water and creates 3.3 tons of landfill waste every year.

in the home

1 reduce

Houses, televisions, meals—as things get bigger, so do the demands on the earth. The average American's ecological footprint (the area of land required to sustain consumption and waste) is more than 25 acres—five times more than what is sustainable globally. We can easily reduce our footprint by avoiding unnecessary consumption. Do you boil enough water for six cups of tea when you want only one? Do you leave the TV or stereo on when you leave the room? Do you throw things away unused? Small acts like these seem insignificant, but collectively they add up and contribute to global warming.

- Less waste
- Less pollution
- Lower carbon emissions
- Save money

Image © APL

reuse 2

Most of what we consume ends up as garbage within months, weeks, days, or even minutes. The United States is the world's most wasteful society, per capita, sending 236 million tons of solid waste to landfills each year. That's an average of 4.5 pounds of trash, per person, per day. By extending the life cycle of products, we can cut down on all that garbage. Try to buy items that are reusable or come in reusable packaging—and make sure you reuse them. Before you throw it away, ask yourself if that wrapping paper can be saved, that container refilled, that pair of shoes repaired, or that machine fixed.

- Less waste
- Less pollution
- Lower carbon emissions
- Save money

Image © APL

3 recycle

Curbside recycling programs mean we now recycle more than 2.35 million tons of glass bottles, 54 billion aluminum cans, and billions more plastic and paper items every year. Recycling reduces landfill and saves resources: Recycling an aluminum can uses only 5 percent of the energy required to make a new one, recycling glass uses 50 percent of the energy, and every ton of paper recycled saves 60 percent of the energy, 17 trees, 7,000 gallons of water, and 60 pounds of air pollution. Yet most people admit to being confused about what they can and can't recycle. Throwing the wrong trash in the recycling bin can contaminate the collection and undermine the viability of recycling efforts. Go to www.earth911.org or www.epa.gov/msw/states to learn more about what your local town or city council recycles.

- Less waste
- Lower carbon emissions
- Less pollution

think of the baby

The average baby will have a minimum of 6,000 diaper changes over 2.5 years. More than 18 billion disposable diapers are used in the United States every year, and 92 percent of those end up in landfills. A staggering 10,000 tons of disposable diapers are thrown into U.S. landfills every day, and the manufacturing of these single-use diapers requires more than 1 million metric tons of wood pulp and 75,000 metric tons of plastic each year.

But environmental and consumer agencies rate reusable cloth diapers as being just as harmful to the environment, depending on what detergent you use and at what temperature you wash them. While disposable diapers create greenhouse gases during the manufacturing process and contribute to landfill, reusable cloth diapers require large amounts of water, energy, and detergents in their washing and drying. Minimize the impact by washing in bulk and line drying, and by choosing disposable brands that are biodegradable and chemical-free.

- Less pollution
- Lower carbon emissions

Image © APL

power shower

L ong showers literally pour resources down the drain, with every minute you linger under the standard showerhead using four to six gallons of water. If you're spending more than five minutes in the shower, you're dawdling. If it's more like 10 minutes, over the course of a year you could be wasting enough water to fill a backyard pool and creating an extra 2,200 pounds of carbon dioxide emissions just from the energy used to heat the water. Reduce your average shower time from 10 minutes to 5 and save more than 4,200 gallons of water each year while cutting your utility bills. Save even more by insulating your hot-water tank, which can eliminate 1,100 pounds of greenhouse gases each year, and setting the thermostat of your water heater to a lower temperature.

- Less water consumption
- Lower carbon emissions
- Lower water and energy bills

Image © APL

stem the flow

N early 25 percent of the water used inside the average home is for showers. A family of 4, each taking a 5-minute shower once a day, uses about 700 gallons of water every week. You can cut that amount in half by using low-flow shower fittings that reduce the flow by 50 percent or more while maintaining ample water pressure. Compared with a standard showerhead, which might use 4 to 6 gallons of water a minute, a water-efficient showerhead can use as little as 1.5 to 2 gallons. After a year of 5-minute showers with a low-flow showerhead, that family of 4 will save up to 20,000 gallons of water as well as the energy needed to heat that water. This in turn reduces greenhouse gas emissions by more than 300 pounds per year and lowers the family's water and energy expenses. Look for efficient plumbing fixtures with the EPA's WaterSense® label.

- Less water consumption
- Lower carbon emissions
- Lower water and energy bills

Image courtesy of Interbath-Australia

7

don't flush it all away

Only three percent of the world's water is fresh, and only a third of that is available for human consumption. Yet Americans are flushing away much of that fresh water every day—literally. Toilets use by far the most water in the home, consuming nearly 40 percent of residential water use. More than 4.8 billion gallons of water are flushed down toilets every day in the USA—that's 9,000 gallons of water per person every year.

An old-style, single-flush toilet can use more than three gallons of water in one flush; modern reduced-flow systems average little more than one gallon. A high-efficiency toilet (HET) gives the same flush with no trade-off in power and can reduce household toilet water use by about 60 percent per year, providing a significant savings on your water bill. If you have an old toilet, reduce its capacity by filling one or more soft-drink bottles with water and placing them in the toilet tank. Look for the HET label when purchasing a new toilet, and fix your leaky toilets: They waste 200 gallons of water every day.

- Less water consumption
- Lower water bills

life's a dish 8

You may pride yourself on a clean kitchen, but the dirty little secret is that sinks and dishwashers account for 15 percent of average household water use and hundreds of pounds of greenhouse gases in hot-water heating. An average automatic dishwasher can use more than 10.5 gallons of water per cycle; an efficient model uses half that. Run the dishwasher only when it is full, and save 10 to 20 gallons of water a day. Doing the dishes the old-fashioned way can use even less, depending on your method. A running faucet wastes more than 2.5 gallons of water a minute, so wash and rinse items together rather than individually. Reduce water flow by half without reducing water pressure by installing inexpensive aerators or flow valves in your faucets, and air-dry your dishes instead of using your dishwasher's drying cycle. Look for the Energy Star® label when buying a new dishwasher; it will significantly lower your utility bills.

- Less water consumption
- Lower greenhouse emissions
- Lower water and energy bills

9 solar flair

Water heating accounts for a large chunk of the greenhouse gas produced by an average household's energy use and 15 percent of a home's energy bills. Every 3.5 gallons of water heated by a conventional electric water heater generate more than two pounds of emissions. There are more efficient ways to heat your household water, including on-demand, tankless, and heat-pump water heaters. One of the most efficient options, however, is to use the sun's rays to heat your water. If you have an electric hot-water heater and an unshaded, south-facing location on your property, consider installing a solar water heater. While the up-front cost is more than a conventional hot-water system, a solar hot-water heater can provide up to 90 percent of a household's hot-water needs and pay for itself within 5 years (you'll only need your existing hot-water heater as an occasional back-up). Moreover, using a solar hot-water heater will eliminate more than 2.5 tons of CO_2 emissions every year, and federal tax credits are available.

- Less water consumption
- Lower greenhouse emissions
- Lower water and energy bills

photo opportunity

10

A part from heating water, solar energy can be harnessed through photovoltaic, or solar, panels that turn light into electricity. In the past, using solar panels to generate energy has not been an economical option for most households. But as the technology improves and production costs come down, solar panels are becoming more viable for the average household. Though it is initially expensive to set up, a photovoltaic system will generate power for thirty years and pay for itself after about eight. Soon solar panels will be cheap and effective enough to pay for themselves within two years. And every kilowatt-hour (kWh) of electricity you avoid using will keep more than 1.5 pounds of CO_2 emissions from polluting the air. To find out if you are eligible for tax credits on solar panels and other energy-efficient home improvements, go to www.dsireusa.org.

- Lower carbon emissions
- Lower energy bills

Image © APL

11 get cozy

Home heating and cooling systems in the U.S. are responsible for releasing 150 million tons of greenhouse gases into the atmosphere every year. Heating and cooling your home accounts for up to 50 percent of your household's energy bills, and air leakage alone accounts for 10 percent or more. As much as 31 percent of heat loss from a house is through the ceiling, walls, and floor. Insulation material made from cellulose, fiberglass, foam, recycled paper, or straw can keep a home cozy and comfortable all year-round, minimizing the need for heaters in winter and air conditioners in summer, and reducing your annual heating and cooling costs by up to 30 percent. Visit www.energysavers.gov to find recommended amounts of home insulation for your climate zone.

- Lower carbon emissions
- Lower energy bills

seal the cracks

12

Every degree of difference in the temperature between the inside and outside of your home can add as much as 10 percent to your heating and cooling expenses. Make the most of the energy you use by trapping the air rather than letting it escape through cracks under doors, between windows, and around floor vents. You can cut greenhouse gas emissions by more than 1,000 pounds a year by using inexpensive seals to caulk and plug cracks and gaps, fitting dampers to fireplaces, blocking unnecessary vents, and weather-stripping all seams. Properly sealing your home costs very little but has huge returns. Check www.energystar.gov to learn more about tax credits for improving your home's efficiency.

• Lower carbon emissions
• Lower energy bills

13
join the fan club

Energy demand peaks on hot summer days as millions of businesses and homes across America turn on their air conditioners. With houses often built for size and view, good passive solar design and orientation are often ignored—and the result is a heat trap (or, in winter, a cold dwelling). A well-designed home should need nothing more energy-intensive than a ceiling fan. Keep windows and curtains closed during the day to block out the heat, then open them at night to let the house cool, and consider a whole-house fan or evaporative cooler for your home. If you must use an air conditioner, set its thermostat to a balmy 81°F rather than a cooler 75°F to reduce energy use by nearly a third, and keep your ceiling fan on—it will improve your air conditioner's efficiency. And remember, installing proper insulation and sealing air leaks in your home will increase energy performance all year long.

- Lower carbon emissions
- Lower energy bills

14

make the most of the day

B anish your daily interior gloom with a natural light source that lets you leave the lights off until the sun goes down. A new generation of highly reflective tubular skylights can bring energy-efficient lighting through tall roofs to rooms deep within two-story and older homes. The same qualities of reflection can be used to light your home in other ways. Walls painted pale colors don't absorb as much light as dark-colored walls. For recessed lights, use lower-watt bulbs with reflector backs.

- Lower carbon emissions
- Lower energy bills

have a 15
light touch

Lighting the average American home generates about two-thirds of a ton of greenhouse gases every year and consumes 20 percent of the average household's electricity bill. Often the lighting is unnecessary—and it is a myth that turning lights on and off uses more electricity than leaving them on. Turn them off if you'll be out of the room for more than a minute. If you find it hard to remember to do this, relatively inexpensive timer controls and daylight or movement sensors can be installed to switch off lights automatically. Dimmers and lamps can also help to reduce unnecessary light use.

- Lower carbon emissions
- Lower energy bills

use bright ideas

The ordinary incandescent light bulb remains the most popular form of home lighting because it is so affordable. But it is also very inefficient, with 95 percent of the electric current being converted into heat, not light. A 20-watt compact fluorescent light (CFL) provides as much light as a 100-watt incandescent bulb and lasts about 8 times longer. Though a CFL will cost approximately 10 times more to buy than an incandescent bulb, over its average life of about 5 years it will use roughly a quarter of the power and save more than 1,400 pounds in greenhouse gases. A lighting store can generally advise on the best product for your needs, with a better-quality CFL lasting up to five times longer than an inexpensive one. Using new lighting technologies can save between half to three quarters of your home lighting energy use—and if every home in America changed just one incandescent bulb to a CFL, we'd save enough energy to light seven million homes and prevent greenhouse gas emissions equal to that of one million cars.

• Lower carbon emissions
• Lower energy bills

blow off the heater

Radiators, or space heaters, consume energy like there's no climate-friendly tomorrow, with a single unit generating more than two pounds of greenhouse gases every hour. Fan heaters are no better; because they rely on convection heating—or directly warming air, which then often escapes through doors, windows, and vents—they are extremely inefficient at heating a large room. Use them sparingly to heat people, not spaces. Natural gas and heat pumps are better alternatives, generating only a third of the emissions of electric radiators, space heaters, and fan heaters. Better still, invite some friends over: Every person in your house generates the same amount of warmth as a 100-watt heater.

- Lower carbon emissions
- Lower energy bills

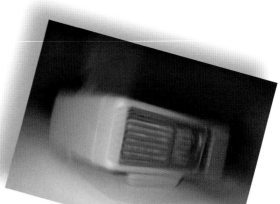

warm yourself, not the environment

18

One of the earliest—and still most efficient—forms of insulation invented was clothing. Long before they became a fashion statement, clothes were helping people survive in a world where there was no artificial heating or air conditioning to keep the temperature at a shopping-mall-constant 75 degrees. Dress appropriately for the weather. Snuggle up in a sweater or thick socks when you feel a chill. If you are still cold, try jumping up and down for a minute, or throw on an extra blanket on winter nights. Temperature variations are a natural part of life, and a fit and healthy body should be comfortable enough without excessive artificial heating or cooling. Every one degree you avoid in external heating by keeping your thermostat down will cut about five percent from your heating bill. In addition, simply by installing a programmable thermostat, you will enjoy significant energy savings, and the thermostat will likely pay for itself in one season.

- Lower carbon emissions
- Lower energy bills

19
step up to the line

An electric clothes dryer generates more than 6.5 pounds of greenhouse gases per load. A solar clothes dryer, better known as a clothesline, generates none. Choose the natural solution whenever you can. If it rains while clothes are hanging, consider it a softening rinse. For those times when you must use a dryer, make sure your washing machine's spin cycle removes as much excess water as possible, dry only full loads, and keep the dryer's lint filter clean so it operates at maximum efficiency. Also, ask your local utility company or government agency about efficient-appliance rebates, and look for the Energy Star® label when purchasing a new washer.

• Lower carbon emissions
• Lower energy bills

Image © APL

it doesn't wash

An average washing machine produces nearly 200 pounds of greenhouse gas emissions in a year, but it's when you choose hot water that cleaning clothes gets really dirty. Between 80 and 85 percent of the energy used to wash clothes comes from heating the water, and a hot-water wash generates 5 times more in greenhouse gases than a cold wash. Choose the cold cycle to save 6.5 pounds in emissions per wash and a significant amount on your utility bills. There's no need to throw in more detergent to compensate, either: The scrubbing action of the washing machine—with front-loading models usually being the most efficient—does most of the work. Nearly three pounds of greenhouse gases are generated in the manufacturing of just 3.5 ounces of detergent, so use only what is needed. Avoid washing clothes that don't need it often—jeans, for instance—and only wash with full loads.

• Lower carbon emissions
• Lower energy bills

cool it with the fridge

21

Nearly 20 percent of U.S. homes have at least two refrigerators. Unless you live in a remote area and need to feed a family the size of the Brady Bunch, this may be a luxury the earth can't afford. The energy output of the refrigerators used in just the U.S. is equal to the output of 60 300-megawatt power plants. Refrigerators in U.S. homes account for 14 percent of the average household's electricity bills, and larger or older-model fridges can use hundreds of dollars in electricity per year, generating nearly 1.5 tons of greenhouse gases. Maximize your fridge's efficiency by making sure the seals work and positioning it in a cool spot. Freezers should be set at 0-5°F, fresh food compartments at about 40°F. When buying a new fridge or freezer, choose one with the Energy Star® label for reduced energy consumption and electricity bills.

- Lower carbon emissions
- Lower energy bills

don't cook up a storm

Cooking in the average U.S. home uses enough energy to generate half a ton of greenhouse gases a year. It's as simple as making a pot of coffee or tea: Every gallon of water boiled produces one pound of emissions. Minimize energy by cooking efficiently: Reuse hot water, put lids on pots, have dishes simmer rather than boil, and don't preheat the oven. There is no standard energy rating for cooking appliances, but bear in mind that a conventional oven will produce a third more greenhouse gas emissions than a convection one, an electric stove produces double the emissions of a gas or microwave oven, and bigger appliances are less efficient than smaller ones. Use a microwave or toaster oven whenever possible.

- Lower carbon emissions
- Lower energy bills

23

get drastic on plastics

The age of plastic walks hand in hand with the age of oil. Our homes are full of objects made from petrochemicals such as polyethylene, polystyrene, polyvinyl chloride, polypropylene, nylon and acrylic. Yet the properties that make plastic so convenient—durability and resistance to degradation—also make it difficult to dispose of. Choose furnishings and household items that will last and can be recycled. Opt for alternatives to plastic made from paper or wood. Use natural plant-based finishes on wood.

- Less pollution
- Healthier home

soft furnishings

Look for natural furnishings that have been or can be recycled—tables made from reclaimed wood, organic cotton curtains, organic wool comforters, beeswax candles, bamboo dishes, hemp bedding, or jute floor coverings. Save resources used in the manufacture of new materials by buying secondhand furniture—old lumber products are much more durable, and far more interesting, than new furniture made from veneer-covered particle board. Or buy new furniture made from recycled materials such as plastics and PET bottles, steel, tires, and recycled-content building materials.

- Less consumption
- Lower carbon emissions
- Less waste

Furniture made with Recopol™ recycled resin shells.
Images courtesy of Wharington International

detox your home

The average home contains more chemicals than an early 20th-century chemistry laboratory: aerosol cans, paints, furniture polish, glues, ammonia-based cleaners, nail polish remover, oils, and battery acid. While the effects of the traces of up to 300 synthetic chemicals that have been found in human bodies are still unclear, for the environment these household chemicals are a proven toxic cocktail when disposed of in landfills or poured down the drain. Check with your local town or city council about its collection days for such chemical wastes. Municipal recycling centers usually accept oil, paints, paint thinners, and cleaners, as do some hardware stores.

• Less pollution
• Healthier home

in the can

Indoor air is three times more polluted than outdoor air, and conventional paints are among the main culprits. Because fossil fuels are the primary ingredients found in paints and varnishes, they give off greenhouse gases, toxic waste, and air pollutants known as volatile organic compounds (VOCs). Buy natural paints and finishes, or water-based paints with low-VOC or zero-VOC content that carry little or no petroleum-based solvents. They are not only better for the environment but also offer an alternative for allergy and asthma sufferers, pregnant women, and young children who are sensitive to certain pollutants. New nontoxic biodegradable paint strippers are also becoming available. And to prolong the life of wood floors, decks, and furniture, use natural wood oils. Unlike with normal polyurethane varnishes, there is no need to sand the wood before applying.

- Lower carbon emissions
- Less pollution
- Healthier home

Image © APL

27

go with the grain

Wood is a perfect renewable and sustainable resource, provided it isn't being clear-cut or harvested in unsustainable fashion. With only 4 percent of our own native, old-growth forests still standing, the U.S. is the world's largest importer and consumer of timber wood and products, bringing in $25 billion each year. But not all of the wood we import is legally or sustainably harvested. As much as seventy percent of timber from Indonesia is destructively or illegally logged, and, closer to home, Mexico has the second highest rate of deforestation in the world, just behind Indonesia. To ensure the wood you are buying has come from a forest managed according to internationally agreed-upon ethical, social, and environmental standards, look for the Forest Stewardship Council (FSC) certified timber label, which is endorsed by organizations including the World Wide Fund for Nature, Friends of the Earth, the Rainforest Alliance, and the Sierra Club.

- Save old-growth forests
- Reduce greenhouse effect
- Support sustainable industry

build, don't destroy

28

A piece of lumber from a local, sustainable plantation has less environmental impact than imported forest timber. Recycled lumber has even less impact. Recycled materials can also save you money and add personality to your environment. You can find bricks, roof tiles, floorings, windows, doors, fireplaces, and fittings at salvage shops and junkyards. Even when recycled materials are not cheaper, you will be adding value to your home and helping to create a market for recycled resources, which in turn will encourage others to recycle. Look for local materials to save money and energy on your own transportation. And whenever possible, buy recycled, secondhand, reclaimed, or waste timber.

- Reduce waste
- Save money

curtain call

29

Cut heat transfer through windows by a third by installing heavy, lined drapes with pelmets or valances. Wooden frames provide better insulation than aluminum. Shade east- and west-facing windows with insulating devices such as blinds or shutters. Cover south-facing windows with suitably angled eaves or awnings that provide shade during summer and light during winter. Choose glass appropriate to orientation and climate. Many Americans live in climates where glass technology can help save energy throughout the year, regardless of which direction their home faces.

• Lower carbon emissions
• Lower energy bills

30

get that glazed look

Windows are the weakest link in a well-insulated home, with a square yard of conventional single-pane glass exposed to direct sun on a hot day generating as much heat as an electric space heater. On a cold day, that same glass will lose more heat than the same area of insulated wall. Double-glazed windows, using two sheets of glass with air or gas sealed between them, are up to twice as expensive but also up to twice as efficient. Use an outer pane that will block unwanted solar radiation and an inner pane that will reduce heat loss from inside. Consider replacing your single-pane windows with low-e coated or Energy Star® windows. If that is not feasible, simply using storm windows can reduce your winter heating costs by 25-50 percent. Energy-efficient windows can lower your heating and cooling costs by up to 35 percent each year, and substantial tax credits are available. To find out more, go to www.efficientwindows.org or www.energystar.gov.

- Lower carbon emissions
- Lower energy bills

31
sleep disorder

That comforting little red or green light emitted by your television, DVD player, stereo system, or computer comes at a price. Every year it is adding about $70 to your electricity bill and creating up to 190 pounds in unnecessary greenhouse gases. Even in standby or "sleep" mode, appliances and home computer equipment can still be operating at up to 40 percent of their full running power. Standby power, which often serves no real function (apart from running a clock), accounts for about 5 percent of U.S. household electricity use and costs consumers $8 billion annually. Rather than leaving your TV or stereo to consume power when it isn't being used, switch it off at the power source. Buy power strips with individual switches to manage several appliances, and turn off the equipment that is not being used. When upgrading home office equipment, look for Energy Star® computers and printers, which use half the electricity of standard office equipment. And consider buying a laptop for your next computer purchase; they use far less energy than desktops.

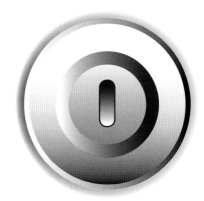

- Lower carbon emissions
- Lower energy bills

32
switch to renewables

Our energy needs will continue to draw on emission-producing, nonrenewable sources such as coal for as long as we care only about what energy source is cheapest to burn. Today, however, there are more ways than ever to use renewable energy at home. Increasing numbers of Americans are taking seriously their responsibility to future generations; they are signing up for renewable power to ensure their energy consumption is not contributing to greenhouse gas emissions. Buy electricity generated by renewable sources, such as solar, wind, and water power, from your local utility. It does cost more, but your money is used to support crucial investment in renewable energy. And buying renewable energy for your home can eliminate as much as 10 tons of carbon dioxide emissions every year. To find out what tax credits you may obtain by using renewable energy, visit: www.dsireusa.org.

- Lower carbon emissions
- Invest in renewable energy

Nationally, 32 percent of residential outdoor water use is due to lawn care. The watering of U.S. landscapes, gardens, and lawns uses nearly 8 billion gallons of water each day.

in the garden

go native

Native animals depend on the plants they have evolved with for food and shelter. Though a few hardy species have adapted to human transformation of their traditional habitat, most native animals find little attraction in imported lawns and rose gardens. Foreign plants will also guzzle water: Running a sprinkler for an hour consumes more than 264 gallons of water, and a standard garden hose uses even more. Native plants will not only attract birds and butterflies but also save the time and expense of daily watering. Growing indigenous plants can save 50 percent of the water typically used to maintain outdoor plants. The U.S. is a country of enormous biodiversity, so check with your local cooperative extension program, nursery, or gardening club about what plants are indigenous to your area.

- Greater biodiversity
- Less water consumption
- Lower water bills

Image © APL

34

multi-story living

Nature finds its balance in a complex network of mutual dependence, with each species providing several useful benefits to other flora and fauna. A tree might feed one animal and shelter another. In natural forests there are several distinct layers of vegetation in which different animals live. Try to replicate this in your garden with a combination of tall trees (pines or sycamores, for instance), smaller trees and tall shrubs (maples, cherry trees, magnolias, hydrangeas, mountain laurel), shrubs (boxwood, junipers), groundcover (grasses and creepers), and a litter layer of leaf matter, fallen branches, logs, and rocks. Use terracotta pipes as substitutes for hollow logs to give small mammals and amphibians a place to hide from predators such as cats and dogs.

- Greater biodiversity
- Less water consumption
- Lower water bills

35
cover up

Between 60 and 70 percent of our treated drinking water is used to water lawns, and the average irrigated home lawn consumes more than 10,000 gallons of water each summer. Conserve water by replacing little-used grass areas, such as your front yard, with a native garden that reduces street noise and increases privacy. Groundcover can thrive where grass does poorly. Putting mulch around plants and on lawns can cut the amount of water lost through evaporation by up to 70 percent. It also limits weed growth and can improve soil conditions. Pine bark mulch can even be used for children's play areas: It is just as safe as grass and requires no watering.

• Lower water consumption
• Lower water bills

shady characters

Plants not only lower greenhouse gas emissions and provide habitats for wildlife; they can also lower home energy costs. Trees with high canopies on the west side of your property will provide shade from the afternoon sun. Shrubs that allow filtered light and breezes are appropriate for more localized shading of east- and west-facing windows. Deciduous trees and vines are useful on your home's southern side, providing foliage to shade against the summer sun while allowing light and warmth during winter. Be sure to shade your air-conditioning unit. Even small plants will help cool your home, through the evaporative process called transpiration, and evergreen trees can be used as a wind barrier, preventing cold winds from reaching your house. You can enjoy significant yearly savings in home heating and cooling costs by landscaping wisely.

- Less energy consumption
- Lower energy bills

Image © APL

37

liquid asset

Imagine watering the garden every time you have a shower or filling the toilet tank whenever you wash your hands. Only about half the water used by the average household needs to be of crystal-clear drinking quality. Water for the toilet and the garden can be recycled from bathtubs, showers, washing machines and the kitchen sink using so-called gray-water recycling systems. The higher content of chemicals like phosphorus and nitrogen in gray water can even be a source of nutrients for plants. A gray-water system can be anything from a tank that collects laundry water to a comprehensive system using plants and micro-organisms to treat water from all household outlets. More and more U.S. cities have implemented municipal water systems for non-potable uses using reclaimed water. Homeowners can do the same, by collecting wash water and using it for other, non-drinkable purposes. Check with your local water authority about rebates for approved gray-water systems.

- Less demand on water supply
- Lower water bills

think tank

The vast majority of the rain run-off in neighborhoods and suburban areas is channeled along gutters and down pipes, straight into the drain—even as the water levels in dams and reservoirs drop. The rain that falls on your roof can be collected in a tank and used on the garden, to fill the pool, or for flushing toilets, doing the laundry, showering, or even drinking. Every 3.5 ounces of rainfall running off the average home roof can provide enough water for several weeks of total household consumption. In fact, using a rain barrel to capture water from a downspout, a 2,000-square-foot home can collect as much as 36,000 gallons of rainwater each year. Several states have rainwater harvesting rebate programs; check with your local water authority.

- Less demand on water supply
- Lower water bills

Image © APL

live under a flight path

39

Birds bring life to a garden, providing color, movement, sound, and useful fertilization and pest-control services. They consume up to half their weight in weed seeds, pests, rodents, and insects each day, and they pollinate flowers and scatter valuable plant seeds. With just 3.7 percent of the U.S. landmass reserved for national parks, and the rapid destruction and degradation of habitat, the backyard garden has become an increasingly valuable potential habitat. Make your garden a desirable destination for many kinds of birds, including migratory songbirds, warblers, woodpeckers, and hummingbirds. Birds like taller trees for roosting and nesting, shrubs for flowers and fruits, and clearings to hunt for seeds and insects. Creating a pesticide-free haven of native plants, hollow logs, or bird houses, along with a supplemental food and fresh-water supply, will put more feathers in your cap. So will keeping your garden predator-free; if you must have a cat, put a bell on its collar and try to keep it inside. Placing decals or window ornaments on clear-glass windows will help prevent collisions.

- Greater biodiversity
- Lower insect population
- Easier gardening

40
make a clean sweep

The latest status symbol of the suburban gardener is the good, old-fashioned—and eco-efficient—broom or rake. Hosing the leaves off the driveway and front walk is out, since it can consume fifty gallons of water every five minutes. So is powering up a two-cycle gas engine to give the lawn a quick blow. Gas-powered lawn mowers, leaf blowers and weed-wackers are not only a source of noise pollution for the neighbors on Sunday mornings; they also spew oily clouds of "debris dust" that are harmful when inhaled and can produce the same amount of air pollutants in an hour as driving a car from Washington, D.C., to Atlanta. Every weekend, 54 million Americans cut their grass with gas-powered mowers, consuming 800 million gallons of gas each year and producing tons of CO_2 emissions. So use a little "elbow grease" instead of harmful pollutants to clean your yard—pull out the trusty rake or broom to rake leaves or sweep off the driveway or patio.

- Lower carbon emissions
- Less noise
- More exercise

Image © APL

grow your own fruit and veggies

41

Even unprocessed foods contain "embodied energy"—the fuel and water consumed in growing, harvesting, transporting, storing, and selling. The farther your food travels to reach your plate, the greater the expended energy and resulting associated greenhouse gas emissions. Having your own fruit and vegetable patch is not only a source of guaranteed fresh organic produce but also very eco-efficient. It's even a great motivation to get outside and get some exercise. A kitchen garden can also be a source of fresh herbs for daily cooking. And if you can't grow your own produce, patronize your local organic vegetable farm or farmer's market—you'll be buying close to home, the produce you purchase will likely be lower in embodied energy, and you'll be helping your local economy.

- Lower carbon emissions
- Better eating
- Lower grocery bills

natural-born killers

Nature regulates itself. A garden dependent upon artificial fertilizers and pesticides is flawed in design. Fertilizers, synthesized from fossil fuels or dug up from phosphate-rich Pacific Islands, can harm native wildlife by promoting the growth of weeds that compete for resources. When washed into lakes, rivers, and streams, they encourage algal growth, killing fish. Pesticides, meanwhile, kill beneficial insects along with pests, can directly harm other species, and pose a threat to humans when they accumulate in the food chain. Fertilizers, when applied improperly, can contaminate ground and surface water. Choose eco-efficient gardening solutions. Native plants will thrive without fertilizer. Insect-eating animals and complementary planting of natural herbicides will reduce the need for pesticides. Use an organic pesticide made out of hot pepper, garlic or chili to control aphids and caterpillars.

- Lower carbon emissions
- Healthier environment
- Less time and money spent on garden

43

get to the roots

Watering systems that deliver water as closely to plant roots as possible, where it's most needed, can reduce evaporation losses by up to 75 percent. The most efficient and low-maintenance watering system is sub-surface drip irrigation, using pipes with small perforations below the soil surface to deliver droplets at a rate of two, four, or eight quarts an hour. A simple sub-surface watering device can be made by cutting the top off a plastic bottle and punching holes in its base. Bury the bottle about eight to 12 inches from the base of the target plant, deep enough so that it will release water about four inches below the surface of the soil. Cycle irrigation systems, bubbler/soaker systems, trickle irrigation systems, and moisture sensors on sprinkler systems are all effective and will also cut water use.

- Less water use
- Lower water bills

from dusk to dawn

44

Image © APL

Plants and soil lose water during the day through evaporation. Watering during the cooler times of the day gives the water a chance to permeate the soil and be absorbed by the roots of the plants. In hot conditions, it is better to water in the evening. In cooler conditions, though, plants can develop fungal growth from water on their leaves overnight, so it is preferable to water early in the morning. Group plants with similar water needs—place thirsty plants together and water them longer, but less often, to encourage deeper roots and increase their drought tolerance. By watering wisely, you can conserve significant amounts of water.

• Less water consumption
• Lower water bills

back to the earth

45

Food scraps and yard clippings make up one quarter of U.S. solid waste. When this organic matter ends up in landfills and decomposes without air, it produces methane, a greenhouse gas 20 times more potent than carbon dioxide. A compost pile or worm farm is a simple but effective way to cleanly convert waste from the kitchen or garden into something productive; you can reduce your yard waste 50-75 percent by composting. Each ton of organic matter you divert from the garbage can will prevent the creation of a third of a ton of greenhouse gases in landfills. Put a pail next to the kitchen sink to collect all food scraps that are not meat or dairy. Leave clippings on the lawn when mowing, and adjust your mower blade higher to cut less off the top of your grass. Yard clippings quickly decompose into nutrient-rich humus, providing up to 25 percent of a lawn's fertilizer needs each year. Compost is also the perfect natural fertilizer, containing all the necessary elements to improve soil structure and microbial activity while retaining soil moisture—and saving precious landfill space.

• Less waste
• Save money on gardening bills

46
use the landscape

Make a feature of the contours in your garden by landscaping for water efficiency. Make the most of water running off sloping, rocky, and paved areas by bordering them with plant beds to absorb the water. Create depressions or ponds where water can collect to seep back into the ground and replenish the water table rather than going straight down the storm-water drain. A permanent pond is a valuable addition to a garden ecosystem, attracting birds and frogs. With the right plantings, it will provide a habitat for native fish, which in turn will eat mosquitoes, other insects, and algae. Or consider Xeriscape™, an increasingly popular, seven-step water-efficient landscaping concept that uses native and climate-adapted plants to achieve beauty and water conservation in yards. A well-planned Xeriscape™ can cut outdoor water consumption by as much as 60 percent.

- Less water waste
- Healthier garden
- Healthier environment

Image © APL

47
pave the way

Roofs, roads, driveways, footpaths, and other water-resistant surfaces stop rainfall from being absorbed into the ground, preventing it from being used by plants and filtered by the soil before entering the water table. The run-off instead flows, untreated, down drains and is dumped directly into rivers, streams, and lakes, flooding the ecosystem with chemicals and other pollutants. The amount of water lost down the storm-water drain is many times that used in the average garden. Lessen the loss by minimizing paved areas and using permeable pavers or paving designs that provide gaps for the rainwater to reach the earth.

• Less water waste
• Healthier garden
• Healthier environment

in the swim

A pool can lose its equivalent volume in water every year, or up to 1,000 gallons every month, as a result of evaporation and over-cleaning. Ensure that your pool is not a huge water drain by filling it from rainwater catchment. Install a backwash storage system, where the water is stored to let chlorine dissipate and then is discharged into a vegetated area of the garden rather than into the storm-water system. Cover your pool when it is not in use to minimize evaporation; pool covers can reduce outdoor and indoor pool heating costs by 50-80 percent. Covering your pool will also help keep it clean, saving on chemical cleaners. Minimize the need for chlorine, which can be harmful to the environment and human health, by using an alternative cleaning system such as an ionizing water purifier. Ask your pool supplier about your options.

- Lower water use
- Healthier environment
- Lower water and cleaning bills

Trees logged from forests account for more than 71 percent of office paper used today, with 8 million tons of copy paper used in the U.S. every year. That's equal to 188 million trees.

at work

49

lunch is on you

Buying your lunch is arguably more water- and energy-efficient than making it yourself, but a home-packed lunch is undoubtedly cheaper and produces less solid waste than fast food. Food courts and other public places can be neglectful in providing recycling bins, with glass bottles, paper napkins, polystyrene plates, plastic cutlery, and food scraps all going into one trash can for disposal in a landfill. Save resources and money by bringing your lunch in a reusable container. Rather than buying new plastic containers, reuse take-out containers before throwing them into the recycling bin. Reuse food wrappers and other plastic packaging rather than buying plastic wrap or aluminum foil.

- Less waste
- Save money

get a mug shot

Coffee has become an indispensable part of the working day, but why not dispense with disposable cups? Life-cycle analysis of the energy and waste from producing, transporting, and disposing of cardboard or polystyrene cups shows the ceramic coffee mug to be far more eco-efficient—even taking into account the water needed to wash it between uses. Over its life span, a mug will be used about 3,000 times, resulting in 30 times less solid waste and 60 times less air pollution than using the equivalent number of cardboard cups. Most take-out coffee shops will be happy to serve your favorite brew in your own favorite mug—after all, it saves their business money—and you are likely to get a slightly larger coffee fix for your effort.

- Less energy use
- Lower carbon emissions
- Less waste

Image © APL

stay in the black

51

Dire warnings against reusing printer ink and toner cartridges contribute to more than 300 million plastic printer cartridges ending up in landfills each year; that's about 8 cartridges every second. Unless the fine print on the warranty states otherwise—many manufacturers will designate a cartridge "for single use only"—there is no reason why a cartridge can't be reused up to four times. You will cut waste and save up to 90 percent on the cost of a new cartridge. Be sure to use a reputable company that will refill or remanufacture your printer cartridges and is prepared to offer a written guarantee against printer damage.

- Less waste
- Save money

re-fill

turn over an old leaf

Despite advances in technology, the paperless office remains a futuristic fantasy, with the typical U.S. worker using a whopping 10,000 sheets of paper—as much paper as is produced by pulping a full-grown tree—each year. Much of this paper comes from native pine forests and is chlorine-bleached, a process that produces toxic dioxins. The simplest way to cut down paper use is by "duplexing," or using both sides. Set the printer and photocopier defaults so that you have to choose *not* to print double-sided. Use the printer's reduction feature to copy two pages onto one sheet of paper. Minimize the potential for waste through paper jams by storing paper in a dry place and loading it into the copier the right way up (it actually *does* make a difference).

- Conserve forests
- Less waste
- Save money

treat it like it grows on trees

Paper comprises nearly three-quarters of office waste, so simple recycling measures can significantly reduce an organization's waste-removal expenses. Print out only what is necessary, and proofread documents carefully on your computer screen to avoid having to print multiple copies. Place a tray on your desk to collect single-side, printed scrap paper and use it for taking notes or in the photocopier or fax machine. Keep a paper-recycling bin under your desk and in communal printing areas, and encourage your colleagues to recycle.

- Conserve forests
- Reduce waste
- Save money

Image © APL

close the loop

A business is not truly recycling unless it buys recycled products. Recycled paper uses up to 90 percent less water and half the energy required to make paper from virgin lumber and produces 36% less greenhouse gas emissions, yet less than 9 percent of the 8 million tons of printing and writing paper used in the U.S. each year is recycled content. The rest comes from chopping down 188 million trees. While recycled papers were once avoided because they looked inferior, could cause copiers to jam, and were unsuitable for archival purposes, it is now often hard to tell the difference, with manufacturers providing recycled paper for virtually all office functions.

• Conserve forests
• Lower carbon emissions
• Lower energy and water use

55

would the last one out...?

The lights in many offices burn long after workers have finished burning the midnight oil. It might make for a pretty skyline, but leaving the lights on, combined with all those computers left on standby, can double a company's energy bill. U.S. commercial buildings alone generate 18 percent of the country's CO_2 emissions. Unnecessary lights also generate unnecessary heat, requiring the air conditioner to work overtime, using even more electricity. The built environment in the U.S. accounts for one-third of all energy consumption and emits the same percentage of pollutants and greenhouse gases, much of that from electricity. Greenhouse gas emissions created by commercial building lights can be reduced significantly—and easily. Just ask your building manager to turn lights off at night or to install movement-activated sensors. Take the initiative by placing reminders near light switches in the area where you work.

PLEASE TURN THE LIGHTS OFF!

- Lower carbon emissions
- Lower energy bills

collect call

You only use your cell phone recharger for a few hours a week, but leaving it plugged into the power outlet means it could be drawing electricity all the time. The same goes for rechargers of other electronic devices, like laptop computers, PDAs, MP3 players, and digital cameras. Unplug them when you're not using them. Phones contain toxic metals—including arsenic, antimony, beryllium, cadmium, copper, lead, nickel and zinc—that do not degrade in the environment, so it is important to recycle them. There are several leading industry recycling programs, among them Recellular Inc.'s www.wirelessrecycling.com and www.charitablerecycling.com, where you can donate your used cell phone and help make a difference, both in the environment and in someone's life. Old mobile phones that still work can be exported to developing countries where they can help bridge the digital divide, be given to domestic nonprofit organizations to assist those in need, or be dismantled for parts and recycled to make other products.

- Lower carbon emissions
- Lower energy bills

start a branch office

57

With many office products emitting potentially harmful toxins, the air in your office may be far from clean. The Environmental Protection Agency (EPA) ranks indoor air pollution as one of the top five environmental public health risks today. The air inside a sealed, energy-efficient building can be 25 to 100 times more polluted than outside. The EPA estimates that the negative impact on workers' health costs the U.S. economy between $17 and $43 billion each year. Indoor plants are natural air filters, absorbing airborne pollutants and radiation from computers while replenishing oxygen levels. Research has found that indoor plants can reduce fatigue, coughs, sore throats, and other cold-related illnesses by up to 30 percent, cutting down on absenteeism. Plants also have a measurable effect on stress levels, helping to keep employees happy and relaxed in the work environment. Improving indoor environments would lower health-care costs, cut sick leave, and improve worker performance, adding up to as much as $30 to $150 billion in annual worker productivity increases.

• Less air pollution
• Healthier workplace
• Lower energy bills

log off and shut down

There are an estimated 55 million office computers in the U.S., and many of them never get switched off, needlessly consuming energy overnight and on weekends. Computers and monitors use more electricity that all other office equipment combined. The average computer left on all day, every day uses nearly 1,000 kilowatts of electricity over the course of a year, producing more than a ton of carbon emissions. In contrast, a computer switched off at the end of the day uses less than 250 kilowatts—and significantly lowers energy bills. Turn off equipment at night, on weekends, and when it is not being used for extended periods of time during the day. If you are away from your computer for shorter periods, put it in sleep mode or enable your PC's power management features, which will reduce energy use to about five percent of full operating power. And buy Energy Star®-compliant office equipment for even more savings.

- Lower carbon emissions
- Lower energy bills

59

loosen your collar

Business attire – particularly the traditional businessman's uniform of long-sleeved shirt, jacket, and tie—is ill-suited to warmer U.S. climates. During the summer months, wearing this more cumbersome clothing means air conditioners have to be cranked up to maintain comfortable conditions in offices as well as in shops and restaurants, using large amounts of energy and contributing to the greenhouse effect. More climate-appropriate work apparel—short-sleeved open-necked shirts, for instance—during summer months, and year-round in warmer climates, means your office air-conditioning won't need to be set to such a low temperature. Every degree higher the thermostat is set will cut up to 20 percent from air-conditioning costs.

- Lower carbon emissions
- Lower energy bills

Image © APL

taken to the cleaners

You think you're picking up fresh, clean work clothes from the dry cleaners…but you might also be getting a sniff of the industry's dirty environmental laundry. Dry cleaners use large amounts of the chemical solvent tetrachloroethylene, a powerful degreasing agent that is also a suspected carcinogen, can aggravate asthma and allergies, and is harmful to the environment. During its production, transport, and use, tetrachloroethylene breaks down into other chemicals—such as the toxin phosgene—and contributes to photochemical smog. Before opting for dry cleaning, consider the merit of a quick, cold-water hand wash or spot cleaning. Look for a cleaning service with "clean and green" processes, including reuse of hangers and garment bags.

- Less pollution
- Healthier environment

61

event horizon

Think green next time you are arranging for the printing of business cards, buying office equipment, or even booking a conference venue. Consider impacts such as greenhouse gas emissions from travel, energy use, water conservation, and waste-minimization policies. The Leadership in Energy and Environmental Design (LEED) Green Building rating system, used by the U.S. Green Building Council, provides a comprehensive indicator of a building's eco-efficiency and sustainability. But don't just stop there: Ask suppliers about their environmental practices and choose businesses that use and provide recycled goods. The purchasing power of a business can help create more demand for recycling and help bring down the cost of recycled goods.

- Lower carbon emissions
- Lower water use
- Less waste

invest in the future

M ost Americans have no idea where or how their pension is invested, but new laws promoting choice in retirement funds give you a chance to put your money where your mouth is. You now have the option to invest your money according to environmentally and socially responsible criteria that reflect your personal values; those criteria might indicate preferred investment in companies with leading environmental credentials (like renewable energy). An industry pension fund, mutual fund, or financial services firm that adheres to socially responsible investment (SRI) practices and has social and environmental screens can offer returns that match or out-perform general funds. Know where your retirement funds are being invested, and choose an ethical investment option.

- Support sustainable industry
- Reap higher returns

call in the auditors

A sk your senior management to commission an environmental audit of your business and the building you work in. A trained environmental consultant will assess the amount of energy, water, and materials your company uses, including where these resources come from and where they end up as waste. An environmental audit can highlight areas of inefficiency and excessive waste, and provide solutions that reduce resource consumption and save money. It is also an effective risk-management tool, helping a business avoid the costly consequences of fines or legal challenges for environmental violations.

• Improve environmental performance
• Improve business efficiency

go public

64

M ore and more, Americans are thinking of the environmental and social implications of their purchasing decisions. More than 92 percent of us say we want businesses to go beyond their historical role of making profits, paying taxes, employing people, and obeying the law; we want businesses to take responsibility for reducing their energy consumption. Our purchasing power can reward those companies doing good things and penalize those doing nothing. Your company benefits from having a comprehensive environmental policy, an environmental officer, and public reporting of its goals and achievements. If your company doesn't do any of these things, ask why not, and volunteer to help establish a committee to oversee, measure, and report on environmental efforts.

- Improve business reputation
- Promote public accountability

One-third of all the trash thrown away in the U.S. is packaging materials. Americans toss more than 75 million tons of packaging and containers—532 pounds for every man, woman, and child—into landfills each year. And less than 40 percent of it is recycled.

shopping

65

less is more

Every day we are bombarded by thousands of advertisements encouraging us to equate quality of life with consumption. You might think of it as retail therapy, but for the environment it is a serious affliction. Every item you buy contains "embodied energy"—the water, fuel, and waste used in its production, packaging, transport, and disposal. Achieving a sustainable lifestyle means buying a bit less of everything. Before you make any purchase, ask yourself if you really need it. In most cases, your life won't be any less full or rich without it, and every dollar you save will reduce your ecological footprint.

- Less energy and water consumption
- Lower carbon emissions
- Less waste
- Save money

Image © APL

borrowing benefits

66

Next time you are in a news agency, bookstore or DVD shop, ask yourself if you really need to own something you will probably only read or watch once. Join the local library and video rental store, and borrow rather than buy books and DVDs. See if you can borrow items you seldom use—power tools and camping gear, for instance—from family and friends. Services usually have a lower environmental impact than goods, so indulge in regular trips to the movie theater rather than spending thousands of dollars on equipment to replicate the experience in your own home.

• Less energy and water consumption
• Lower carbon emissions
• Less waste
• Save money

buy second-hand

67

Apart from food, clothes shopping has the highest environmental impact of all consumer activities, with about 39,600 gallons of water used in the production and transport of the new clothes bought by the average American household each year. Resist artificially created fashion cycles and step out in your own recycled style. Rather than buying a new pair of jeans that have undergone an industrial process to give them that worn look, just buy a pre-worn pair. The garments you can find in vintage clothing shops are often better than new items and cost a fraction of the price.

• Less energy and water consumption
• Lower carbon emissions
• Less waste
• Save money

lasting glory

RECHARGE

Buying the cheapest toaster, washing machine, DVD player, light bulb, or battery is rarely the most cost-effective option for your wallet—or the environment. Such products are usually made with inferior parts that quickly wear out and cannot be replaced—a dubious design practice known as "planned obsolescence" that ensures continued demand at the expense of valuable resources and unnecessary waste. Cheap designs also usually preclude effective recycling of components. Investing in well-designed, more durable items that can be repaired, upgraded, reused, and recycled saves money in the longer term. Rechargeable batteries, for instance, can be recharged hundreds of times, with each charge lasting up to twice as long as a disposable battery, making them an economical option for power-hungry devices like digital cameras.

- Less energy and water consumption
- Lower carbon emissions
- Less waste
- Save money

pack it in

69

From soft drinks and sugar packets to cheese singles and shrink-wrapped cucumbers, the amount of packaging we consume continues to proliferate. Each year our country throws away more than 75 million tons of packaging and containers—that's a staggering 532 pounds for every man, woman, and child—and less than 40 percent of it is recycled. All plastics marked 1 to 7 are theoretically recyclable, though some cities and towns only recover 1 (polyethylene terephthalate), 2 (high-density polyethylene), and 3 (polyvinyl chloride). Check which plastics your community recycles and avoid purchasing packaging that will end up in a landfill. Buy in bulk where you can, and choose fresh produce from local markets over plastic-wrapped supermarket goods. Avoid excessive packaging: Pay for the product, not the package. Choose reusable packages—and reuse them.

- Less energy and water consumption
- Lower carbon emissions
- Less waste
- Save money

refuse plastic bags

Plastic bags are one of the most ubiquitous items on the planet. Americans go through 380 billion of them each year, and about 100 million of those are just for shopping (at a cost to retailers of $4 billion). Their use can typically be measured in minutes—the time it takes you to get home from the store—but the bags can last for hundreds of years. Although most go into landfills—only 0.6 percent are recycled—an estimated 100 million are let loose in the wild. Those airborne bags, often called "urban tumbleweed," clog sewers, gutters, and waterways, entangle birds, and are swallowed by hundreds of thousands of whales, turtles, and other marine life. The oceans are full of tiny fragments of plastic slowly working their way into the food chain. So take a reusable bag when you go shopping, and say "no" to plastic bags offered at shops and grocery stores.

- Less energy and water consumption
- Lower carbon emissions
- Less waste
- Healthier environment

true green

71

balance your diet

About 132 gallons of water are required to produce a pound of potatoes, 505 gallons to produce a pound of rice, and a whopping 26,400 gallons to produce one pound of grain-fed beef. Cattle, sheep, and other livestock account for a large portion of agricultural water use, which in turn accounts for 87 percent of America's total freshwater consumption. Agriculture is the second largest source of greenhouse gases after the energy sector, and livestock are responsible for much of the soil erosion in the U.S., damaging the environment by trampling native vegetation and compacting soil with their hooves. Choose to eat less resource-intensive meat such as chicken and turkey, and eat more grains, fruits, and vegetables.

- Less energy and water consumption
- Lower carbon emissions

72
eat it

While wasting resources on excessive food packaging is bad enough, wasting food, which has far greater embodied energy and water, is worse. We eat more than we need—as the nation's expanding waistline attests—and buy more than we eat. A shocking 27 percent of all food produced each year in the U.S. for human consumption is tossed in the garbage. That amounts to 48 million tons of food wasted each year, or about 163 pounds of food for every man, woman, and child. More than $31 billion worth of edible food ends up in landfills every year—about $600 worth for the average family of four. What's worse, approximately 49 million people could have been fed by that wasted food. Take to heart that old advice to eat everything on your plate, make the effort to reuse leftovers, and buy only what you need.

- Less energy and water consumption
- Lower carbon emissions

Image © APL

73

there's a catch

Seafood has long been considered an essential part of a healthy diet, and, more recently, science has identified the health benefits of fish oil, but rising consumption is putting increased pressure on fish stocks. Since the early 1990s, the number of species threatened by overfishing has risen significantly, and some 39 million tons of "by-catch," or unwanted fish, are killed and tossed back into the sea because of inadequate fishing practices. At the current rate of fishing, many species could be eliminated within the next quarter-century. Be conscious of the seafood you're consuming. Consider choosing sustainably farmed varieties, such as wild Alaskan salmon or Pacific albacore tuna. If you need more information on the seafood offered by your local market or restaurant, the Seafood Choices Alliance, which publishes "The Fish List," provides extensive resources on specific species, sustainable farming practices, and current environmental conditions.

- Save endangered species
- Protect biodiversity

74

edible packaging

For eco-efficient and fully biodegradable food packaging, choose fresh fruits and vegetables. Their skins are more-than-adequate protection for transporting them from the store to your home and can be recycled as compost to feed your garden. Raw fruits and vegetables, as well as legumes and nuts, also use less energy and water than refined and processed food. Eating more fruits and vegetables and unprocessed food is healthier, reducing the risk of obesity, allergies, heart disease, cancer, and other ailments. So rediscover the pleasure of biting into a crunchy apple rather than a chocolate bar. You'll feel and look better for it.

- Less energy and water consumption
- Lower carbon emissions
- Less waste
- Save money

75 trade fair

Americans consume one-fifth of the world's coffee supply, more than any other country on the planet. Yet precious little of the proceeds from all that coffee goes back to the local coffee farmers toiling in developing nations. Globalized food production often results in the export of exploitative practices to Third World countries, including the abandonment of traditional farming practices, the clearing of rain forests to create more arable land, and the planting of single-variety cash crops dependent upon artificial fertilizers and pesticides. By contrast, Fair Trade-branded products are sourced directly from local cooperatives, putting more money in the pockets of the growers, who can then invest in more sustainable farming as well as their children's education. Look for Fair Trade products, including coffee, tea, cocoa, and chocolate at your supermarket, organic food store, or coffee shop.

- Promote sustainable agriculture
- Reduce inequality
- Healthier eating

as nature intended

Organic produce, grown without the use of fossil fuel-based fertilizers, synthetic pesticides, or genetic modification, is becoming increasingly popular as we become more concerned about the health risks of chemical-laden food. In contrast to non-organic farming, where nutrients are applied to the soil in a soluble form, organic farming focuses on the underlying health of the soil, with plants taking up nutrients that are released naturally from humus by microbes. The environmental dividend is greater biodiversity at all levels of the food chain. Organic produce often contains significantly higher concentrations of essential vitamins and minerals. Plus, it usually tastes better.

• Lower energy and water use
• Lower carbon emissions
• Healthier environment
• Healthier eating

star quality

Energy rating labels are mandatory on fridges, freezers, air-conditioners, washing machines, and dishwashers. The rating is determined by energy efficiency as well as product size, so a bigger appliance with a high rating might still use more energy than a smaller appliance with a lower rating. Some machines, like computers, have energy-saving features that must first be activated, so make sure you do so. Look for goods carrying the Energy Star® label, which indicates the environmental performance of a product from a whole-product-life perspective.

• Lower carbon emissions
• Lower energy bills

clean living

The very products and processes used to keep indoor environments clean may also contribute to indoor pollution, or "sick building syndrome." Apart from poisoning through ingestion—mostly by small children—studies show that in certain conditions, many everyday household cleaners and air fresheners emit toxic contaminants at levels that pose risks to our health. Common symptoms include eye, nose, and throat irritation, as well as headaches, dizziness, and fatigue. Long-term or cumulative environmental consequences, such as contamination of surface and ground water, may also occur. Clean up with a micro-fiber cleaning cloth, warm water, a dash of natural soap, and good, old-fashioned elbow grease. Ingredients like vinegar, borax, lemon juice and baking soda are effective on tougher stains.

• Healthier environment
• Healthier home

79 get personal

The average person uses between 10 and 45 pounds of soaps, toiletries and cosmetics each year. Hair sprays, shaving creams, shampoos, deodorants, perfumes, and other personal-care products contain an array of active ingredients that are dangerous in high doses. Many cosmetics and toiletries have been found to contain chemicals that are either known carcinogens or are simply untested—of the many thousands of synthetic chemicals used in everyday household personal-care items, less than 20 percent have been tested for acute effects and less than 10 percent for chronic, reproductive, or mutagenic effects. While the small amounts of these ingredients that you absorb through the skin may not be enough to cause any noticeable harm, when washed down the drain these ingredients accumulate at levels equal to agrichemicals (the array of toxic and hazardous chemical products used in enormous quantities in agriculture). There is no universal safety test for personal-care products, and using virtually any such product with any chemical ingredient may pose some level of risk. For your own health and that of the environment, look for natural alternatives.

- Healthier environment
- Healthier home

80

moral fiber

While synthetic textiles such as nylon, polyester and Lycra are produced from fossil fuels, opting for natural fibers is not the clear-cut environmental choice you might think. Cotton is the world's most chemical-intensive crop, requiring 10 to 18 applications of herbicides, insecticides, and fungicides, as well as more than 3,800 gallons of water, per pound produced. One-quarter of global pesticide use is on cotton crops. Wool, meanwhile, requires a staggering 22,400 gallons of water per pound and is associated with soil compaction caused by sheep hooves, pesticide applications to manage lice and flies, habitat loss due to demand for fertilizer-dependent pastures, and potent methane emissions. Chemical-free organic cotton, linen, wool, and hemp are the best alternatives.

- Less water consumption
- Less pollution
- Healthier environment

going the 81
extra mile

Don't let your food be more traveled than you. The average supermarket item is transported hundreds, if not thousands, of miles. The amount of food shipped between nations has grown four-fold over the past 40 years, as supermarket giants source cheap products from poorer nations with lower labor costs or from richer nations that dump subsidized agricultural produce on the international market. But the farther the food travels, the greater the associated carbon emissions. Reduce your food miles by buying local produce at the local farmers' market. Check labels to see how far the food has come, and choose seasonal fruits and vegetables that aren't from half a world away.

- Lower carbon emissions
- Healthier economy

check the bottom line

It is easy to see the environmental problems caused by heavy industry such as mining and energy production, but the truth is that all companies affect the environment through their business practices and decisions. A bank or insurance company, for instance, can make a difference through policies that give better rates or premiums for more eco-efficient homes or cars. Check the environmental credentials of every business you give money to and compare its efforts with its competitors. If it can't demonstrate any environmental commitment—or worse, turns out to be actively lobbying against initiatives like the Kyoto Protocol—take your business elsewhere.

- Healthier environment
- Greater customer satisfaction

The U.S. transportation system is the largest in the world, and it accounts for one-third of America's greenhouse gas emissions—more than 515 million tons of CO_2 each year. That's nearly 70 percent of the oil consumed in the U.S. and more than we as a nation produce.

travel

hoof it

The average family car travels about 15,000 miles a year, generating about 5.8 tons of greenhouse gas pollution and costing more than $2,500 in gasoline alone. In addition, vehicle exhaust contributes to smog, which kills about 30,000 Americans every year—more than the number killed in traffic accidents. Create a cleaner environment by getting in some of those 10,000 steps you need to take every day to stay fit. Walk to local shops rather than taking the car to a distant shopping center. Go for a hike rather than a drive in the country. Every gallon of gas you avoid using saves 1.3 pounds in greenhouse gas pollution.

- Lower carbon emissions
- Less air pollutants
- Save money
- More exercise

get on your bike

Image courtesy of the Australian Greenhouse Office, Department of the Environment and Heritage

84

Greenhouse emissions from transportation—America's second largest contribution to global warming—are rising faster than in any other sector as we increasingly rely on cars to get to work, collect the groceries, drop the kids off at school, pick up take-out food, and return the videos. Half of all car trips are less than 3 miles—a distance research has shown can be covered just as quickly on a bike once traffic and parking are taken into account. Cycling is eco-efficient and fun. Rather than driving to the gym, cycle to work one day a week. A 12-mile trip once a week saves about half a ton of greenhouse gas emissions over a year.

- Lower carbon emissions
- Less air pollution
- Save money
- More exercise

become a passenger

85

Even though America has only 30 percent of the world's cars, we consume nearly half of the global daily fuel consumption—that's 9 million barrels of oil each day. More than 200 million cars and light trucks currently travel U.S. roads, consuming 40 percent of U.S. oil and emitting 20 percent of U.S CO_2 pollutants. And while six in ten Americans have public transportation available to them, only ten percent use it with some frequency, and just four percent use it as their primary means of getting to work (even fewer walk or cycle). Over a year, taking a bus instead of driving a car for a typical ten-mile commute saves two tons of carbon dioxide emissions and reduces air pollutants such as carbon monoxide, nitrogen oxide, particulate matter, volatile organic compounds, and benzene. Using public transport leads to improved and more efficient services as well as quicker travel times due to lower traffic congestion. Or consider a car-sharing service. You'll save money, drive less, and spare the atmosphere from tons of toxic emissions.

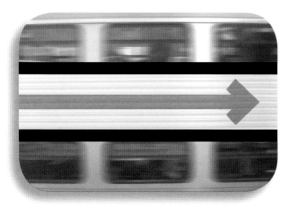

- Lower carbon emissions
- Less air pollution
- Save money

think small

America's love of cars is no secret—and it shows no signs of stopping. U.S. cars and light trucks are driven farther each year and consume 15 percent more fuel than our driving counterparts in other countries. That adds up to 346 million tons of carbon emissions each year—enough to circle the planet twice. The U.S. automotive industry has made great strides in its efforts to reduce pollution. But our obsession with bigger, heavier cars, particularly so-called sport utility vehicles—many of which never see terrain bumpier than the shopping center parking lot—has offset improvements in fuel efficiency. A large SUV in city traffic will get as little as 12 mpg and will emit more than 13 tons of CO_2 annually, compared with about 20 mpg and 8 tons of CO_2 for a traditional 6-cylinder family wagon and 25 mpg and 6.5 tons of CO_2 for a smaller 4-seat sedan. The average car in its lifetime will emit 70 tons of CO_2; the average SUV will emit 100 tons. Driving a gas-guzzling SUV instead of an average new car for 1 year would waste more energy than leaving a color TV turned on for 28 years or a refrigerator door open for 6 years. A more fuel-efficient car can save between $300-$700 a year in fuel costs and more than 2 tons of greenhouse gases.

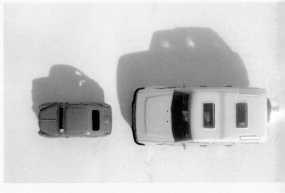

- Lower carbon emissions
- Less air pollution
- Save money

emission control

It takes about 35 medium-sized trees to offset the carbon emissions released into the atmosphere by the average SUV driving 15,000 miles each year. For the cost of about one tank of gas, you can join organizations such as the National Arbor Day Foundation and American Forests to help support tree planting and reforestation efforts that will help reduce the amount of greenhouse gases released into the air by your car. For $30, Trees for the Future will plant 300 trees for you, to offset the amount of greenhouse gases your car will give off in its lifetime. Or go the extra mile and calculate the greenhouse gases produced by your particular vehicle. American Forests has a "Climate Change Calculator" that will tell you how many trees you must plant to absorb the CO_2 your car has put into the environment. Trees also help prevent soil erosion, improve water quality, and provide habitats for native species.

- Offset total vehicle emissions
- Improve biodiversity

choose a hybrid

A family-sized hybrid-engine car is more fuel-efficient than even the smallest conventional models. Using both an internal combustion engine and an electric motor, a hybrid car charges electric batteries from both its internal-combustion engine and the kinetic energy dissipated during deceleration. Its electric motor helps to accelerate the car, takes over while cruising or when idling, and otherwise acts as a generator. When the batteries run low, the gas engine kicks in, recharging them in the process. A hybrid engine uses around 1 gallon of fuel for every 60 miles traveled, and burns fuel more cleanly, with up to half the emissions of a comparably sized car. A family driving a hybrid car will spend only $800-$1,500 per year in fuel compared to the $2,500-$2,800 a family will spend driving a conventional car. Federal tax credits are available for some hybrid and alternative-fuel vehicles.

- Lower carbon emissions
- Less air pollution
- Save money

electric battery + gas =

save

89
soft-pedal

How you drive can make a big difference in the amount of fuel you consume. By accelerating slowly, driving at moderate speed, and avoiding the need for hard braking, you can dramatically increase the mileage you get from a tank of fuel. Avoid high speeds: At 75 mph, your car uses 15 percent more fuel than it does cruising at 65 mph. A car engine also produces about 40 percent more emissions when cold, so avoid short trips. Plan your journey to combine multiple errands, and, if possible, avoid rush-hour traffic. Removing just one 20-mile trip each week from your routine can prevent more than 1,200 pounds of greenhouse gases a year. If you are stopping your car for more than ten seconds, turn off your engine. An idling engine consumes more fuel and produces more CO_2 emissions that just restarting your car. And keeping your car well maintained and your tires inflated to the correct pressure level also helps fuel efficiency.

- Lower carbon emissions
- Less air pollution
- Save money

90
small strokes

Three in every four car trips involves transporting a single occupant—the driver. We'd save eight billion gallons of gas each year if every commuter car in the U.S. carried just one more person. But if you aren't taking a passenger, a small scooter or electric bicycle is an obvious choice—particularly as an alternative to a household's second car—when it's too far or inconvenient to walk, cycle, or use public transport. A larger scooter with a 250cc engine will use less than 1 gallon of gas per 50 miles and a 50cc engine as little as half a gallon. Scooters are also cheaper and easier to park. With the thousands of dollars you save in running costs and carbon emissions, you could catch a cab when it's raining or rent a car for a weekend away.

- Lower carbon emissions
- Less air pollution
- Save money

Image © APL

fuel around

Not all fuels are created equal. When burned, each gallon of unleaded gas creates nearly 20 pounds of carbon dioxide. High-octane fuel, which contains up to one-third less sulfur than regular unleaded gas, provides more engine power, more efficient consumption, and cleaner exhaust emissions. Look for gas blended with bio-fuels made from renewable or recycled sources such as ethanol (made from cereals and sugarcane) and bio-diesel (derived from vegetable oils or animal waste). A ten percent ethanol blend produces a third of the air pollutants found in conventional gas. Buy established bio-fuel brands to be sure they have been correctly blended at the refinery.

- Lower carbon emissions
- Less air pollution
- Healthier environment

the sky's
the limit

92

Probably the single worst thing you can do for the environment is to jet around for business or vacations. Air travel produces about as much carbon dioxide as each passenger driving a personal car the same distance—and aircraft emissions, released high in the atmosphere, have a greenhouse effect three times greater than road vehicle emissions. A single, one-way coast-to-coast trip will dump an additional ton of CO_2 and other greenhouse gases, per passenger, into the atmosphere. That's double the emissions you'd release by driving cross-country in a carbon-producing SUV. The best way to help offset the greenhouse gases your travels create is to support programs, such as renewable energy and energy-saving projects, which will help reduce the CO_2 emissions created by such air travel. Remember that you can further offset carbon pollutants by supporting programs that plant trees. And plan a holiday that's as much about the journey as the destination. Wherever you go, be an ecotourist: take nothing but pictures and leave nothing but footprints.

- Lower carbon emissions
- Less air pollution
- Healthier environment

An editor who receives 10 letters on the same issue takes interest; a politician who receives 100 letters takes notice.

in the community

93

sit down and be counted

An editor who receives 10 letters on the same issue takes interest, a politician who receives 100 takes notice, and a chief executive who receives 1,000 letters takes action. Decision-makers hear from the public far less than you would expect, so they presume each letter represents the views of dozens, or even hundreds, of people who never got around to writing. And contrary to popular opinion, public servants often crave greater community feedback. Commit yourself to writing one letter a month, or even a week. Rather than complain, make positive suggestions. Even a single letter published in the local newspaper can be the catalyst for thought and action.

• Change policies
• Change opinions

sharing is caring

Sharing builds relationships and communities, and reduces your ecological footprint. Share your tools, yours toys, your time—and, most importantly, your home. The growth in single households is the single greatest contributor to the growth in household carbon emissions, with people living alone using more energy and resources than those who live with others. A person living alone produces double the waste of someone sharing with three others, and single households also waste resources by duplicating the need for household goods such as washing machines, sofas, microwave ovens, TVs, vacuum cleaners, and cooking utensils.

- Lower carbon emissions
- Less waste
- Less consumption

95

money talks

If you've bought shares in companies directly, consider selling them and reinvesting the money through a socially responsible investment (SRI) fund that reflects your social and environmental values. The increasing financial clout of these SRI funds is an important driver of more sustainable business practices. Their investments provide crucial dollars for research and development, which leads to more efficient use of renewable or recycled resources. Cheaper prices lead to greater demand, which in turn creates economies of scale and even lower production costs. If you only have a savings account, choose a community-based bank or credit union that provides affordable loans to not-for-profit enterprises like child-care or health-related services.

- Reward environmental performance
- Expand the market for sustainable products

support local enterprise

S mall local businesses are not only the backbone of the national economy, they're the lifeblood of vibrant local communities, too. They create more local jobs, are responsible for more money being reinvested back into the local community, and provide more scope for local producers than large chain retailers. Local trade also reduces the energy used in transportation. Rather than doing all your grocery shopping in a one-stop supermarket, spend a little extra time visiting the local grocer, butcher, and baker. The experience will be more enjoyable, and you'll often find yourself surprised at how much better value the product is compared with the mass-retailed wares of the industry giants.

- Support the local economy
- Lower transport costs
- Lower carbon emissions and packaging waste

97

show cooperative spirit

Cooperatives can be found in many walks of life—from organic farmers' markets and utility cooperatives to housing collectives, credit unions, community radio stations, and internet service providers. Whether producers, consumers, or workers, they pool assets for the shared gain of their members, creating local investment, services, and jobs (reducing the need for commuting). What's more, they are based on shared values of self-help, equality, and self-responsibility. Whatever profits they make (if that is their intention) are also more likely to be reinvested in the local community. Cooperatives can create products and services that other companies might not regard as profitable enough, thereby building a market for more sustainable business practices. Worldwide, an estimated 800 million people are members of local cooperatives.

- Support the local economy
- Encourage sustainable business

get involved

The concentration of media coverage on national and international issues and the global magnitude of the environmental challenges we face can leave us feeling powerless and thinking that any action is futile. But the problems we need to tackle are usually close to home, in our very neighborhoods if not our own homes. There are thousands of local groups making a difference across America—planting trees, recycling goods, turning used cooking oil into bio-diesel, cleaning beaches, or promoting fair trade. If you can't find a group that fits your interests, gather a few like-minded souls and start your own. That's how many of the country's most well-known and successful community and environmental groups began—as a local, grass-roots campaign.

- Improve the local environment
- Offset carbon emissions
- Encourage sustainable business

let it be a lesson

The thrifty lessons we were taught as children—"waste not, want not," for instance—turn out to be deep-seated environmental wisdom. The habits we learn early are usually the ones we keep. Combine your community involvement with a learning experience by supporting (or starting) a sustainability initiative at your local school. It could be a permaculture garden (which creates a sustainable habitat by duplicating nature's patterns), with the fresh produce being used in the school cafeteria; a "walking bus" program, using a network of adult volunteers to lead neighborhood children to and from school by foot; an organic food cooperative or a secondhand school uniform and book store; or a recycling drive to raise money for the school or a charitable cause. Teach your children social responsibility at any early age; it will likely grow along with them.

- Improve the local environment
- Offset carbon emissions
- Encourage a sustainable future

Artwork by Lachlan Chang

the secret of happiness

100

Image © APL

Though many aspects of our culture encourage us to believe that the important things in life are wealth and fame, extensive academic research across different countries and cultures has shown that happiness has little to do with spending money. The most contented people are those who spend their time helping others and contributing to their community. From peeling potatoes in a soup kitchen to working pro bono, volunteering gives you an opportunity not only to increase the social capital of the nation—the cornerstone of its economic prosperity—but also to find personal fulfillment in an activity that doesn't revolve around consuming resources. Reduce. Reuse. Recycle.

- Make the world a better place
- Reduce your environmental footprint
- Be happier and healthier

web resources

Footprint Calculators	Carbon Footprint	www.carbonfund.org
	Carbon Calculator	www.americanforests.org
	Ecological Footprint	www.myfootprint.org
	Energy Calculators	www.eere.energy.gov/consumer/calculators
Water	Water-Savings Tips	www.h2ouse.net
	WaterSense	www.epa.gov/watersense
	Water Use It Wisely	www.wateruseitwisely.com
Energy	Alliance to Save Energy	www.ase.org/consumers
	Energy Savers	www.energysavers.gov
	The Power is in Your Hands	www.powerisinyourhands.org
	U.S. Dept. of Energy, Office of Energy Efficiency and Renewable Energy	www.eere.energy.gov
Home and Garden	Coalition Against the Misuse of Pesticides	www.beyondpesticides.org
	Eco-Friendly Paints	www.eartheasy.com
	EPA, Indigenous Plants Landscaping	www.epa.gov/greenacres
	Gardens Alive!	www.gardensalive.com
	Gardener's Supply	www.gardeners.com
	USDA, Home Conservation Advice	www.nrcs.usda.gov/feature/backyard
New Homes	Build-e, Eco-Friendly Houses	www.build-e.com
	Certified Forests Products Council	www.certifiedwood.org
	Environmentally Construction Outfitters	www.environproducts.com
	Environmental Home Center	www.environmentalhomecenter.com
	No. American Insulation Manuf.'s Assoc.	www.naima.org
	U.S. Green Building Council	www.usgbc.org
Directory Services	EnviroLink Network	www.envirolink.org
	Green Pages Co-op	www.greenpages.org
	National Environmental Directory	www.environmentaldirectory.net
	Green Living Source for the Consumer	www.thegreenguide.com
At Work	Computer Recycling	www.computerrecyclingdirectory.com
	Conservatree	www.conservatree.com
	Office Footprint Calculator	www.thegreenoffice.com
	Reduce.Org	www.reduce.org
	Reducing Office Waste	www.filebankinc.com/reports/reduction_tips.html
	The Real Earth, Inc.	www.treeco.com

Environmental Labeling	Energy-Star® Rating System	www.energystar.gov
	NSF International (Certification System)	www.nsf.org
Local Recycling	Nationwide Local Recycling Programs	www.earth911.org
	NSF's Recycling Guide	www.nsf.org/consumer/recycling
Computer Recycling	Earth 911	www.earth911.org
	Electronics Industries Alliance	www.eiae.org
Phone Recycling	Collective Good International	www.collectivegood.com
	The Charitable Recycling Program	www.charitablerecycling.com
	Wireless recycling	www.wirelessrecycling.com
	Wireless Foundation	www.wirelessfoundation.org
Investment	Social Funds/SRI World Group, Inc.	www.socialfunds.com
	Social Investment Forum	www.socialinvest.org
Food	Eartheasy	www.eartheasy.com
	Equal Exchange	www.equalexchange.com
	Green Restaurant Association	www.dinegreen.com
	Seafood Choices Alliance	www.seafoodchoices.com
	TransFair USA	www.transfairusa.org
	Whole Foods Market	www.wholefoods.com
Shopping	Co-op America, National Green Pages	www.coopamerica.org/pubs/greenpages/
	Earth Animal	www.earthanimal.com
	Ecomall—Environmental Shopping Center	www.ecomall.com
	Global Exchange/Fair Trade	www.globalexchange.org
	One Percent for the Planet	www.onepercentfortheplanet.org
	Professional Wet-cleaning Network	www.tpwn.net
	Responsible Shopper	www.responsibleshopper.org
	Reusable Shopping Bags	www.reuseablebags.com
Transport	Center for Climate Change & Environmental Forecasting	www.climate.dot.gov
	Car Information—Mileage, Hybrids	www.fueleconomy.gov
	Electric Vehicle Assoc. of America	www.evaa.org
	Environmental Guide to Cars and Trucks	www.greenercars.com
	Transportation Almanac—Energy, Pollution	www.bicycleuniverse.info

For the Kids	Bobbie Big Foot	www.kidsfootprint.org
	Cool Kids for A Cool Climate	www.coolkidsforacoolclimate.com
	Earth Kids 911	www.earthkids911.org
	Water Use Calculator	www.ga.water.usgs.gov/edu/sq3.html

Sustainable Lifestyles	American Forests	www.americanforests.org
	Earth Easy	www.eartheasy.com
	Earth 911	www.earth911.org
	National Arbor Day Foundation	www.arborday.org
	National Parks Conservation Association	www.npca.org
	Trees for the Future	www.treesftf.org

Advocacy Groups and Organizations	American Council for An Energy-Efficient Economy	www.aceee.org
	Blue Ocean Institute	www.blueocean.org
	Children's Health Environmental Coalition	www.checnet.org
	Earthshare	www.earthshare.org
	Environmental Defense Fund	www.environmentaldefense.org
	Environmental Protection Agency	www.epa.gov
	Environmental Working Group	www.ewg.org
	Friends of the Earth	www.foe.org
	Green Peace USA	www.greenpeace.org
	Natural Resources Defense Council	www.nrdc.org
	Rainforest Action Network	www.ran.org
	Redefining Progress	www.rprogress.org
	Rocky Mountain Institute	www.rmi.org
	Stop Global Warming	www.stopglobalwarming.org
	The Conservation Fund	www.conservationfund.org
	The Nature Conservancy	www.nature.org
	The Ocean Conservancy	www.oceanconservancy.org
	Worldwatch Institute	www.worldwatch.org
	World Resources Institute	www.wri.org
	World Wildlife Fund	www.wwf.org.

Media	E/The Environmental Magazine	www.emagazine.com
	Earth Policy Institute	www.earth-policy.org
	Environmental Issues Newsletter	www.environment.about.com
	Environmental Health News	www.environmentalhealthnews.org
	Environmental News Network	www.enn.com
	Grist Magazine	www.grist.org
	Tree Hugger, Online Magazine	www.treehugger.com

authors

Kim McKay (right) is the co-founder and deputy chairwoman of Clean Up the World and Clean Up Australia. She is an international social marketing consultant who counts National Geographic among her clients.

Jenny Bonnin is a director of Clean Up Australia and Clean Up the World. She and Kim are partners in the social marketing firm Momentum2. Jenny has two children and lives with her partner and extended family. Both Kim and Jenny live in Sydney, Australia.

acknowledgments

Special thanks to Ian Kiernan for his inspiration, friendship, unstinting good humor and commitment to making a difference; Kathy Stark who has helped from the beginning with Clean Up the World and has researched information sources and made the changes for the US edition; Marian Kyte for the inspired design and imaginative photographic research; and Tim Wallace for his dedication, insight and creative editing. Also thanks to Katie Patrick (Green Pages Australia) for her original research and advice; Michele Wood for her initial research; Lee McLachlan, Tania Baxter, and Molly Harriss Olson.

The authors also wish to thank Nina Hoffman, Kevin Mulroy, and Lauren Pruneski at National Geographic Books for their support, enthusiasm, and encouragement.

Huge thanks also to the Clean Up Australia and Clean Up the World teams who work tirelessly to spread the word; and special thanks to our fellow directors Victor Kelly and John Buttle who were there at the beginning; also Giselle McHugh, Emily O'Neil and Melanie Booth who helped start a global movement.

Photograph: Marc Stanley, titomedia

Clean up the world

about Clean Up the World

Clean Up the World, the international outreach campaign of Clean Up Australia, was, co-founded by *True Green* author Kim McKay and Ian Kiernan, AO— legendary yachtsman and 1994 Australian of the Year.

In partnership with the United Nations Environment Programme (UNEP), Clean Up the World annually attracts more than 35 million volunteers who join community-led initiatives to clean up, fix up, and conserve their local environment.

Fifteen years after its launch, the campaign has become a successful action program that spans more than 120 countries, encouraging communities to take control of their own destiny by improving the health of their community and environment.

Global activities include waste collection, education campaigns, environmental concerts, creative competitions, and exhibitions on improving water quality, planting trees, minimizing waste, reducing green house gas emissions, and establishing recycling centers.

Participants range from whole countries (e.g. Australia and Poland), community and environmental groups, schools, government departments, businesses, consumer and industry organizations, to sponsors and dedicated individuals who either work independently in their local communities or with other groups in a coordinated effort at a regional or national level.

Visit the Clean Up the World website to find out how your community, company, or organization can become involved: **www.cleanuptheworld.org**

"For more than 15 years Clean Up the World has empowered individuals to take care of our environment. The work of our volunteers has made and will continue to make significant inroads, but now it's time to move to the next stage and address the significant environmental threats that face us today in the key areas of climate change, waste and water."

Ian Kiernan, AO
Chairman & Founder, Clean Up the World

"Globally, we need to learn from the
custom and culture of our Indigenous
people and tread lightly on the earth—
in the belief that we don't own
the land, we belong to it.
It's up to us."

Ian Kiernan, AO
Chairman & Founder
Clean Up the World &
Clean Up Australia